DISCARD

# BEAN

David M. Schwartz is an award-winning author of children's books, on a wide variety of topics, loved by children around the world. Dwight Kuhn's scientific expertise and artful eye work together with the camera to capture the awesome wonder of the natural world.

**Please visit our web site at: www.garethstevens.com**
**For a free color catalog describing Gareth Stevens Publishing's list of high-quality books and multimedia programs, call 1-800-542-2595 (USA) or 1-800-461-9120 (Canada).**
**Gareth Stevens Publishing's Fax: (414) 332-3567.**

**Library of Congress Cataloging-in-Publication Data**

Schwartz, David M.
    Bean / by David M. Schwartz; photographs by Dwight Kuhn. — North American ed.
      p. cm. — (Life cycles: a springboards into science series)
    Includes bibliographical references and index.
    ISBN 0-8368-2970-0 (lib. bdg.)
    1. Common bean—Life cycles—Juvenile literature. [1. Beans.] I. Kuhn, Dwight, ill.
II. Title.
QK495.L52S38   2001
583'.74—dc21                        2001031466

This North American edition first published in 2001 by
**Gareth Stevens Publishing**
A World Almanac Education Group Company
330 West Olive Street, Suite 100
Milwaukee, WI 53212 USA

First published in the United States in 1999 by Creative Teaching Press, Inc., P.O. Box 2723, Huntington Beach, CA 92647-0723. Text © 1999 by David M. Schwartz; photographs © 1999 by Dwight Kuhn. Additional end matter © 2001 by Gareth Stevens, Inc.

Gareth Stevens editor: Mary Dykstra

Printed in the United States of America

1 2 3 4 5 6 7 8 9 05 04 03 02 01

# BEAN

by David M. Schwartz
photographs by Dwight Kuhn

A SPRINGBOARDS INTO
SCIENCE
SERIES

**Gareth Stevens Publishing**
A WORLD ALMANAC EDUCATION GROUP COMPANY

What would a garden be like without beans? People grow beans because they are good to eat, but beans also have another purpose. They contain seeds that will grow into new bean plants.

Pods are the fruit of bean plants. If you run your fingers up and down a bean pod, you will feel bumps. Each bump is a seed. If you open a pod, you can see the seeds. When you put a bean seed in soil and give it water, something magical will happen.

Inside each bean
seed is a tiny
plant, or embryo.
The seed also has
two large white
parts called
seed leaves. These
special leaves are
full of food for
the embryo.

In moist soil, the seed slowly swells with water, and the embryo's tiny root begins to grow. The root breaks through the outer shell, or seed coat, and pushes downward into the soil. The seed has germinated.

A pale green shoot
also begins to grow.
As the shoot pushes
upward, it forces
apart the seed leaves
and the seed coat.
The seed coat falls
to the ground.

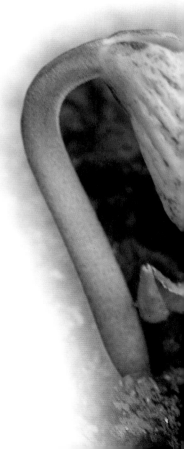

The shoot has two
new leaves that spread
open to face the Sun.

As the shoot, or seedling, keeps growing, more leaves form. The leaves make food for the whole plant. To make food, the leaves use water from the soil, light from the Sun, and carbon dioxide from the air.

Beetles and other insect pests eat bean leaves. If the insects do not eat too many leaves, the plant will be strong enough to keep growing. When the plant is big enough, flower buds form and quickly open into flowers.

Other insects, such as bees, visit the flowers to sip their nectar. A yellow dust called pollen sticks to the insects' bodies. When an insect moves to another flower, the pollen rubs off and fertilizes the tiny eggs deep inside the flower.

The eggs begin to grow into new seeds, and a bean pod forms around them.

The flowers wither and drop off, but the pods keep growing. The seeds inside the pods get bigger and bigger. The fruit and seeds are good to eat, so people pick the pods. If they save some seeds and plant them, the seeds will grow into new bean plants.

Can you put these steps in the
life cycle of a bean in order?

**buds:** small swellings or growths on a stem or a branch that contain leaves or flowers that are not fully developed.

**carbon dioxide:** a colorless, odorless gas that animals breathe out of their lungs and plants take in to make food.

**embryo:** the first stage of growth in the life of a plant or an animal.

**fertilizes:** brings male and female cells together so a new plant or animal can grow.

**fruit:** the part of a plant, such as a pod or a berry, that contains the plant's seeds.

**germinated:** started to grow.

**moist:** a little wet; damp.

**nectar:** the sweet liquid in flowers that many insects and birds like to drink.

**pests:** insects and other small animals that are annoying or cause trouble or damage.

**pods:** shells or casings that grow on a plant and have the seeds of the plant inside them.

**seed leaves:** the first leaves produced by a plant embryo after a seed starts to grow.

**shoot (n):** the part of a new plant that breaks through the soil when a seed begins to grow.

**swells:** gets bigger or bulges because there is something pushing outward from inside.

**wither:** become limp and shrivel up or dry out from not getting enough water.

# ACTIVITIES

## Now You Seed It

If you open a can of green beans, you will find seeds inside the pods. Where could you find other seeds in your kitchen? With an adult's help, go on a seed hunt. Look for other canned foods that have seeds. Then find foods with seeds in the refrigerator or in the cupboards.

## Pick a Packet

Go to a garden center or another store that sells vegetable seeds. How many different types of green bean seeds can you find there? Why are there several varieties? Look closely at a seed packet to learn about growing a bean plant. Do the words and pictures on the packet tell you how to plant the seeds? Now design your own seed packet. Be sure to include all the information a gardener will need to know.

## Inside the Bean Scene

Soak some dried lima bean seeds in water overnight. Take the seeds out of the water and look at them through a magnifying glass. With an adult's help, cut a few of the seeds in half so that you can look closely at what is inside the seed coat. Can you find the beginnings of the root and the seed leaves like the ones shown in this book?

## Sprout It Out!

Line the inside of a jar with wet paper towels. Put a little water in the bottom of the jar, then carefully place a few bean seeds between the sides of the jar and the wet paper towels. Set the jar on a windowsill for about a week. Add water, as needed, to keep the paper towels wet. Check the seeds every day to watch for changes. What do you see happening?

**More Books to Read**

*From Seed to Plant.* Gail Gibbons (Holiday House)

*Grow It Again. Kids Can Do It (series).* Elizabeth MacLeod (Kids Can Press)

*How a Seed Grows.* Helene J. Jordon (Econo-Clad Books)

*Jack's Garden.* Henry Cole (Greenwillow)

*A Kid's Guide to How Vegetables Grow. Digging in the Dirt (series).* Patricia Ayers (Rosen)

*One Bean.* Anne Rockwell (Walker & Co. Library)

**Videos**

*Look What I Grew: Windowsill Gardens.* (Intervideo)

*The Magic School Bus Goes to Seed.* (Scholastic)

*Wonders of Growing Plants.* (Churchill Media)

**Web Sites**

kidscience.about.com/kids/kidscience/library/weekly/aa091900a.htm

tqjunior.thinkquest.org/3715/

www.urbanext.uiuc.edu/gpe/case5/case5.html

Some web sites stay current longer than others. For additional web sites, use a good search engine to locate the following topics: *beans, gardening, plants,* and *seeds.*

# INDEX